YOUR KNOWLEDGE HAS VALUE

Shah Muhammad Butt

Cloud Centric Real Time Mobile Learning System for Computer Science

GRIN Verlag

Bibliografische Information der Deutschen Nationalbibliothek:

Die Deutsche Bibliothek verzeichnet diese Publikation in der Deutschen National-
bibliografie; detaillierte bibliografische Daten sind im Internet über http://dnb.d-
nb.de/ abrufbar.

Imprint:

Copyright © 2014 GRIN Verlag GmbH
Druck und Bindung: Books on Demand GmbH, Norderstedt Germany
ISBN: 978-3-656-63427-0

This book at GRIN:

http://www.grin.com/en/e-book/271355/cloud-centric-real-time-mobile-learning-
system-for-computer-science

GRIN - Your knowledge has value

Der GRIN Verlag publiziert seit 1998 wissenschaftliche Arbeiten von Studenten, Hochschullehrern und anderen Akademikern als eBook und gedrucktes Buch. Die Verlagswebsite www.grin.com ist die ideale Plattform zur Veröffentlichung von Hausarbeiten, Abschlussarbeiten, wissenschaftlichen Aufsätzen, Dissertationen und Fachbüchern.

Visit us on the internet:

http://www.grin.com/

http://www.facebook.com/grincom

http://www.twitter.com/grin_com

Cloud Centric Real Time Mobile Learning System for Computer Science

Shah Muhammad Butt
Department of Computer Science,
Shah Abdul Latif University, Khairpur.

Abstract-The emerging concept of cloud computing, and advancement in mobile devices enabled with Sensor technologies transforms many areas of modern day living. The main advantage of the cloud centric environment is that this technology reduces the infrastructure and software cost and License for all. Mobile learning (M-learning) is considering more effective system of providing study materials to learners anywhere anytime. It is necessary to develop the robust contents delivery mechanism for learning which can be accessed from internet enabled mobile devices. In this paper, we discuss the impact of cloud computing and supporting technologies to acquire education in broader dimension for the students over the country. We believe cloud based mobile learning will surely enhance the current educational system and improve the education quality at low cost.

Keywords: Cloud Computing, Mobile Learning Cost effective Cloud, Mobile Cloud

II.   Introduction: The propagation of mobile devices that comprise smart phones, ipads, laptops, PDAs, tablets and kindles inside an ecosystem that is gradually online, creates the acceptance of program driven through technology conceivable[1][2][3]. These gadgets are sharing daily lives of more and more personal of all ages. Mobile learning will more obscure the peripheries of conventional learning areas. Smart economical mobile devices will build learning accessible anytime and anywhere. Mobile devices can be used to access information and multimedia related to locations, study through mobile apps, access online class materials, and communicate through email, text, and social networks[4]. The Cloud based applications are consider as Internet oriented services on shared and manageable infrastructure (Figure: 1). Due to its Elasticity and the customized software and the Infrastructure and the cost effectiveness; the technology rapidly adopted for business, educational and other discipline[5].

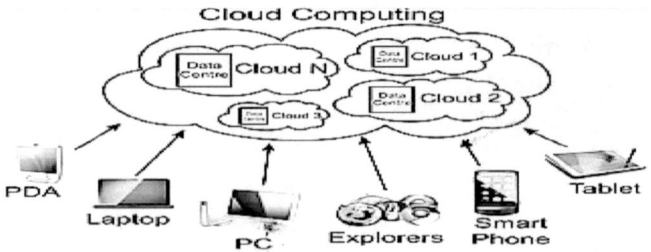

Figure-1: Cloud Computing

III. Background: Providing reliable computational / storage services for learning and research have stand facing immense [6] challenges. Cloud Computing provide a cost effective and elastic IT solutions for learning environment [7]. Cloud computing is a paradigm shifting that provides applications, data storage and processing power over the Internet. The three service models of the cloud are SaaS, PaaS and IaaS (Figure: 2); all are enable users to run and store applications and data online. Software as a Service is the effortless method to extend software services where applications middleware are processed and accessed over the Internet. Many online word processor, spreadsheets and presentation packages are made available for public freely or offered on Pay-As-You-Go (PAYGo) model; which offers data processing and storage services. These renowned public cloud providers such as Google, Microsoft and IBM offer various SaaS applications for educational purposes. Platform as a Service (PaaS) provides customized tools and development environment that enables users to design, develop, run and maintain their specific applications online; App Engine by Google, Azure by Microsoft, Amazon AWS (Amazon Web Service) and, IBM Tivoli offering PaaS cloud Services. The

Infrastructure as a Service (IaaS) model provides infrastructure services; the organizations acquire components such as computing power, Network Services and storage capacity with complete control over the entire IT infrastructure including the hosting environment and replications [8].Clouds promote [9] the delivery of enterprise class Network services provide following services, i.e. (virtualization, network intelligence, Storage and a robust ecosystem). The mobile platform is greatly impacted by this technology as well. Mobile Cloud Computing, speculates that the cloud will rapidly turn out to be a backbone in the mobile world, ultimately occurring the leading technique in which mobile applications operate. Users of Smartphones mobiles have Internet access through 3G services and WiFi networking and the interoperability between networks [10]. The advance functionality of smartphones allows several applications can provide context-aware information about the real time movement, location, and reception on the mobile and store, produce, and share multimedia data using built-in cameras and microphones. According to ITU [11] at the end of 2013, over 5 billion mobile subscriptions in the world. According to Hamblen, smartphones accounted for more than 500 million 2013.

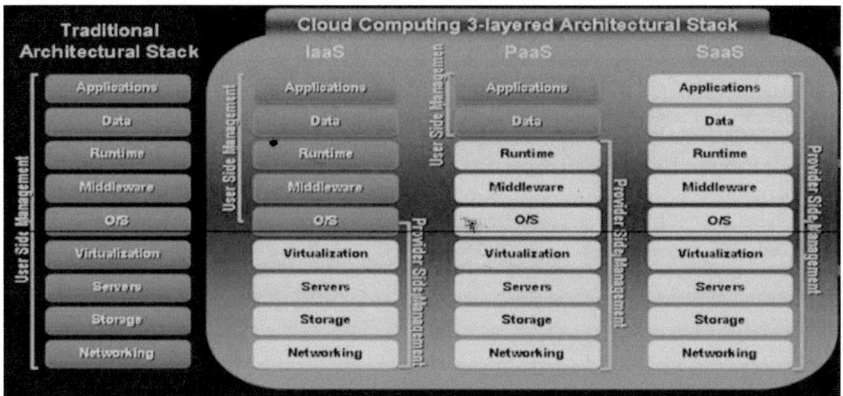

Figure 2: Traditional I.T Architecture vs. Cloud Computing Architecture.

Mobile Cloud Computing [12] supporting various applications containing M-Commerce, M-Learning, M-healthcare, M-Banking and M-Gaming and many other areas. Mobile applications obtained significant global mobile market share several mobile applications have grasped the benefit of Mobile Cloud Computing; [13] the main objective of cloud based M-learning system for education is that the learners can obtain the data from the widely accessible pool of shared resources freely or on Pay As Use; the applications offer users with improved services in terms of storage size, better processing speed, and battery backup.

IV. PROPOSED MOBILE LEARNING SYSTEM FOR EDUCTION: M-Learning is the enhanced approach of E-learning considering the wide spread usage and development of mobile technologies; the method to deliver educational contents through digital media, two recognized methodologies used for E-Learning; i.e. synchronous and asynchronous learning. Synchronous or real time learning (Figure:3) have advantage is that; with limited bandwidth archived and stored contents are accessed and managed without any delay[14].

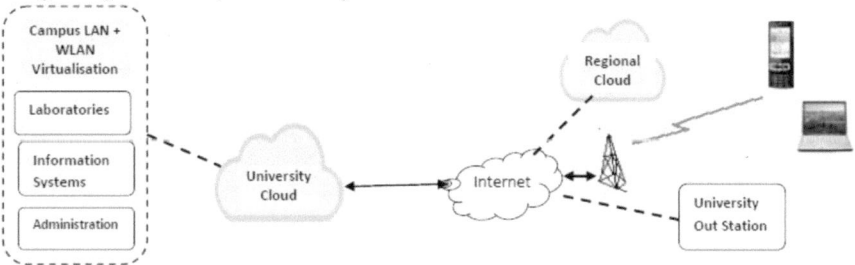

Figure-3: Mobile Cloud learning for Higher Education

To achieve synchronous communication data association and sharing of applications is required. Through this mode interacting with learning contents (text, audio and media files), student attendance, feedback and quarries and class scheduling tools. The student's web based registration required for the downloading of mobile application via GPRS, 3G and Wi-Fi connection. The application access the Computer Science contents cloud for browsing of particular topics of the user [15]. Cloud based contents can be documents or audio video files, tutorials and lectures buffered on the students mobile device; simultaneously student assessment and results could be accessible worldwide (Figure: 4). Furthermore; context-aware documents and tutorials managed over the cloud storage[16]. The advantages of Cloud centric Mobile education over traditional E-learning is; it is cost-effective on demand services, [17] easily implemented, broader network access, rapid elasticity in software services and instant updates, interoperability between hardware and software's of different venders on single platform. Cloud automatically manages data and network load balancing, so that data and services are accessible during peak hours. Cloud-based web applications can deliver learning console for students and teachers accessing 24/7 with applications and data.

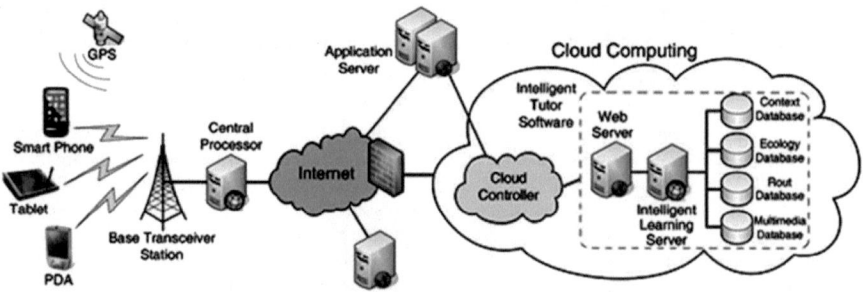

Figure 4: Cloud Centric Mobile Learning System for Computer Science

V. Conclusion and Future work: The University students of current era are more interacting with digital and mobile devices such as smartphones, laptops and tablets. Considering the current trend, it is the time and available technology, introducing the cost-effective learning system is obviously not a big challenge of the current age. Mobile learning techniques will be the essential part in near future, facilitate students with enhanced learning components, designed for particular discipline. The strategies may be adopted for the development of online course contents particularly for the students of Computer Science. Using synchronous (real time) learning method academia and teachers could able to obtain conventional form of computer science learning into digital and smart learning. This paper reinforces the concept of utilizing mobile learning as an interface to increase learning practices for Computer Science and in other fields of higher education. The proposed system will enhance advancements in the current educational system in order to benefit from the technology for teachers and students.

VI. References:

[1] P. Gupta and S. Gupta, "Mobile Cloud Computing : The Future of Cloud," vol. 1, no. 3, pp. 134–145, 2012.

[2] H. T. Dinh, C. Lee, D. Niyato, and P. Wang, "A Survey of Mobile Cloud Computing : Architecture , Applications , and Approaches," no. Cc, pp. 1–38, 2011.

[3] N. Fernando, S. W. Loke, and W. Rahayu, "Mobile cloud computing: A survey," *Futur. Gener. Comput. Syst.*, vol. 29, no. 1, pp. 84–106, Jan. 2013.

[4] N. M. Rao, "Cloud Computing Through Mobile-Learning," vol. 1, no. 6, 2010.

[5] T. Soyata, H. Ba, and W. Heinzelman, "Accelerating Mobile-Cloud Computing : A Survey," 2013.

[6] W. Alsaggaf, M. Hamilton, and J. Harland, "Mobile Learning in Computer Science Lectures," *Int. J. e-Education, e-Business, e-Management e-Learning*, vol. 2, no. 6, 2012.

[7] I. A. Alshalabi, S. Hamada, and K. Elleithy, "Research Learning Theories that Entail M-Learning Education Related to Computer Science and Engineering Courses," vol. 2, no. March, pp. 88–95, 2013.

[8] D. Fesehaye, Y. Gao, K. Nahrstedt, and G. Wang, "Impact of Cloudlets on Interactive Mobile Cloud Applications," *2012 IEEE 16th Int. Enterp. Distrib. Object Comput. Conf.*, pp. 123–132, Sep. 2012.

[9] K. Kim, S. Lee, and P. Congdon, "On Cloud-Centric Network Architecture for Multi-Dimensional Mobility Categories and Subject Descriptors," pp. 1–6, 2011.

[10] E. Zaharescu, "Enhanced Virtual E-Learning Environments Using Cloud Computing Architectures," vol. 2, pp. 31–41, 2012.

[11] Z. Sanaei, S. Abolfazli, A. Gani, and S. Member, "Heterogeneity in Mobile Cloud Computing : Taxonomy and Open Challenges," pp. 1–24, 2012.

[12] C. Savill-smith, "Mobile learning anytime everywhere Mobile learning anytime everywhere," 2004.

[13] S. Pisey, P. L. Ramteke, and B. R. Burghate, "Mobile learning exploring the challenges and opportunities of distance education," vol. 2, no. 3, pp. 19–23, 2012.

[14] P. Shu, F. Liu, H. Jin, M. Chen, F. Wen, Y. Qu, and B. Li, "eTime: Energy-efficient transmission between cloud and mobile devices," *2013 Proc. IEEE INFOCOM*, pp. 195–199, Apr. 2013.

[15] H. Qi and A. Gani, "Research on Mobile Cloud Computing : Review , Trend and Perspectives," 2014.

[16] I. A. Alshalabi and K. Ellcithy, "E FFECTIVE M- LEARNING D ESIGN S TRATEGIES FOR C OMPUTER S CIENCE AND E NGINEERING," 2012.

[17] S. Kitanov and D. Davcev, "Mobile Learning in Mobile Cloud Computing Environment," vol. 8, no. December, pp. 27–39, 2012.